Easy Multiplication Table

Dedication:-

To all the global children.

About The Book:

This book is very scientific and easy for the children to understand and memorise multiplication table. All the tables are sum-up to three small tables which can be memorised very easily. Things are produced in a scientific and new way. It's more a research work than a book by the author. Children need to momorise 3 small tables only. Clear instructions are given to the teachers for proper teaching method in a practical way. A must buy for all the English medium students not only in India but also in abroad. Price is also affordable.

Name:..

Standard:..

School:..

Contents

Heading	Page no	Heading	Page no
Dedication	1	Exercise-7A	29
Contents	2	Exercise-7B	30
Author's Note	3	Exercise-7C	31
Publisher's Note	3	Lesson-8	32
Responsibilities of Teachers	4	Exercise-8	34
Study Plan	6	Lesson-9	35
Few Words To Students	7	Exercise-9	36
Lesson-1A	8	Lesson-10	37
Lesson-1B	9	Exercise-10	38
Lesson-1C	10	Lesson-11	39
Exercise-1A	12	Traditional Multiplication Chart.	40
Writing Method	13	Table-1	42
Exercise-1B	14	Exercise-11	43
Exercise-1C	16	Lesson-12	43
Lesson-2	17	Table-2	44
Exercise-2	19	Exercise-12	45
Lesson-3	20	Lesson-13	45
Exercise-3	21	Table-3	46
Lesson-4	22	Exercise-13A	47
Exercise-4	23	Exercise-13B	48
Lesson-5	24	Figure	49
Exercise-5	25	Notes	50
Lesson-6	26		
Exercise-6	27		
Lesson-7	28		

Author's Note

As such there is no great technical errors in the process of learning English multiplication table but this book is aimed to reduce the burden of memorisation from the head of children. Childrens are childrens only. They cannot cope up with the huge amount of memorization prosedure during their study. In this book the multiplication table has been shortened and multiplication table of 1 to 20 is included and is divided into three tables. The tables are named as Table-1, 2 and 3. The old format is also given from 0 to 9 only. '0' is given special importance in this book. So, the learning of numbers starts from '0' and not from '1'as in the traditional system. So, each set of ten's goes like 0-9,10-19,20-29.....and so on and not as 1-10, 11-20....as in the old method. The multiplication table of 0 and 1 is not included in the table because it is to be taught to the students for anything is multiplied by zero(0) the product is always zero(0) and anything is multiplied by '1' the product remains the same as the number itself. It is also to be put in the brain of the students that anything or any number is multiplied by '10' is nothing but the '0' to be added on the right side of that number only. For example, if you are multiplying 13 with 10 the product will be 130 or if you multiply 15 with 10 the product will be 150. So that is very easy. That is why the multiplication table of 0,1,10 and 20 are not included. Only the multiplication tables from 2 to 9 and then from 11 to 19 are included. Hope this book will be of immense help to the students all over the world and in India also. For any question or suggestion contact the following id or whatsapp.

Thanks and Regards
Amitabha

e-mail ; amitabhamj@gmail.com

Whatsapp: +919832594357 (no call)

Publisher's Note

Respected author Swami Amitabha Anand has taken research on how to reduce the burden of memorization from the children and to remove the fear of mathematics from their mind and in an attempt to do so he has reduced the whole multiplication table in very short tabular form. We are very pleased to publish his research work in front of you all parents, teachers and students. Hope this will be of great help to all the students of English medium schools.

Responsibilities of Teachers

Respected teachers, this book is meant for both teachers and students alike. Kindly note that without the help of a teacher a children can not learn multiplication table or mathematics. This book will definitely help them to form a strong base of mathematics. In this book the full multiplication table is just reduced to 3 small tables. Childrens need to memorize only those three tables and nothing else. So the resultant product has been reduced to at least two third or one third of the present volume and in every table you are to pronounce the smaller number first and then the greater number and then the product. That is how the table is being made.

While learning multiplication table many things we learn twice. For example, in the multiplication table of 3 also we learn 3 za 6 equals to 18 and in the multiplication table of 6 also we learn 6 za 3 equals to 18. But in this book you are to learn only once. That is 3 za 6 eauals to 18 only need to be memorised. More so the multiplication table of 0,1,10 and 20 are not included in this book for definite reason. It is the responsibility of the teachers to put it in the brain of the students that anything is multiplied by '0', the product is always '0'. So there is no need of learning multiplication table of '0'. Similarly it is to be put in the brain of the students that anything is multiplied by '10' the product will be just a 0 on the right side of the number. For example, if 15 is multiplied by 10 then it will be 150 that is '0' is put on the right side of the number 15. Same is the case if I take 48. If 48 is multiplied by 10 then it becomes 480 that is '0' is put on the right side of 48. Very easy. Multiplying with any two digit number having '0' on the unit position is very Easy. Suppose if 25 is multiplied by 20. 25 x 20 = 500, that is 25 x2 =50 and then '0' will be put on the right side of the product so it becomes 500. So the table of 10 or 20 is not included. It is the responsibility of the teachers to put in the brain of the students of why they need to memorise the multiplication table. This helps them to solve bigger mathematical problems in future life. In this book the learning of number starts from '0' and not from '1' unlike in the traditional method. In the traditional method students first learns about '0' while learning 10. This seems illogical. So, Lesson -1B will contain from '0' to '9' and not from 1-10. So, each set of 10's will go like 10-19, 20-29, 30-39.....and so on. The whole book is divided into few lessons, few exercises and few tables. Before students understand two digit number they should have a basic idea of what addition is. The teachers are requested to teach the children with colourful plastic blocks or wooden blocks about numbers from 0 to 9. For example you take 3 blocks in hands and count in front of the student and tell them that these are three blocks. Then as 5 blocks or 7 blocks and in this way from 0 to 9 blocks you can take and teach them what is meant by 0 to 9. Also make them pronounce from '0' to '9'. After that only this book is to be given in their hand . And now to teach the students what addition is take help of the same colourful blocks. Take three blocks in one box and two blocks in another box. Now put them together and show them how many blocks are there. In this case it is definitely 5. Teach them that this is addition. This process is being elaborately explained in Lesson-2. Now before they learn multiplication table also the idea of multiplication should be very clear to the students and this process is being elaborately explained in Lesson-9. Then only the teacher should go for helping them memorizing multiplication tables. 3 multiplication tables are there like table 1,2 and 3. Memorisation of table 1 and 2 is must for all the students and 3rd table is optional. It is strongly suggested to memorise that also.

For any kind of suggestion or queries respected teachers and guardians can contact the following email id and WhatsApp number

Thanks and regards

Amitabha

e-mail : amitabhamj@gmail.com

WhatsApp : 9832594357 (no call)

Warning : No portion of this book be reproduced in any form either printing or electronic or be shared via social media in any form. Any violation to this rule will be entertained by legal consequences.

However reproduction of parts of the book be made with prior permission of the author in written through WhatsApp or e-mail that may incurr some price value.

Study Plan

The whole book is divided into few lessons and the respective exercises. Students will start to learn from lesson- 1A which contains pictures only. They are to count and learn from their teachers how many objects are there. It is not important which object is there; it is important how many objects are there. It's not important whether elephant is there or banana is there but it is important how many elephants or how many banana is there. After that they will learn to read the English numbers with their spellings. Then they will apply the same in the subsequent picture tables. They will count and they will learn the numbers. After that they will have one exercise. They will recognise the newly learnt numbers and their spellings. After that they will learn to write by practicing writing method. Another exercise is there where they will count and write the numbers themselves. In the next exercise they will write and read both the numbers and the spellings. That is how the lesson-1 is being constructed. After the lesson-1 is completed, they will learn how to add different numbers, that means they will learn addition. In this chapter teachers are advised to teach them addition by practical method and instructions are given in the lesson-2 clearly. After that they will solve the exercise-2 with the help of teachers. From lesson-3 to lesson-7 they will learn from 10 to 99 and solve the exercises with the help of the teachers. After lesson 7 is completed teachers will guide the students to repeat once from lesson-1 to lesson-7 again. Then solve exercise-7B and exercise-7C. After that teachers are instructed to teach the students the process of multiplication with the help of colourful blocks or stones and when they learn it properly they'll be guided to the lesson-11 for memorizing the multiplication table. There are three multiplication tables. Table-1 is to be memorized first then table 2 and table 3. However memorizing table 3 is optional . Table-1 and Table-2 must be memorized. It may take some 6 months for an ordinary student to learn this book but if someone takes some one year or more, it doesn't matter. Childrens are childrens only. They are not to be pressurized for learning. Teachers can make their own study plan to teach the book. It is not possible for children to study continuously everyday. There are occasions and leaves and holidays. Everyday at least 30 to 40 minutes study is suggested. This book is not particularly meant for any particular class. But parents and teachers are advised to teach this book to the primary students. Schools may include this book in their syllabus for the 1st and 2nd standard or even for 3rd or 4th.

Thanks.

Few Words To Students

Dear Students, in this book the whole multiplication tables have been shortened to 3 small tables only. This will help you in solving bigger mathematical problems when you grow up. So you are advised to memorize at least Table-1 and Table-2 and if you memorise Table-3, that will be a great help to you.

Thank you.

Wish you all success in your life.

Yours Affectionately

Beloved Author.

Lesson – 1A

Let us learn to count.(With the help of teachers)

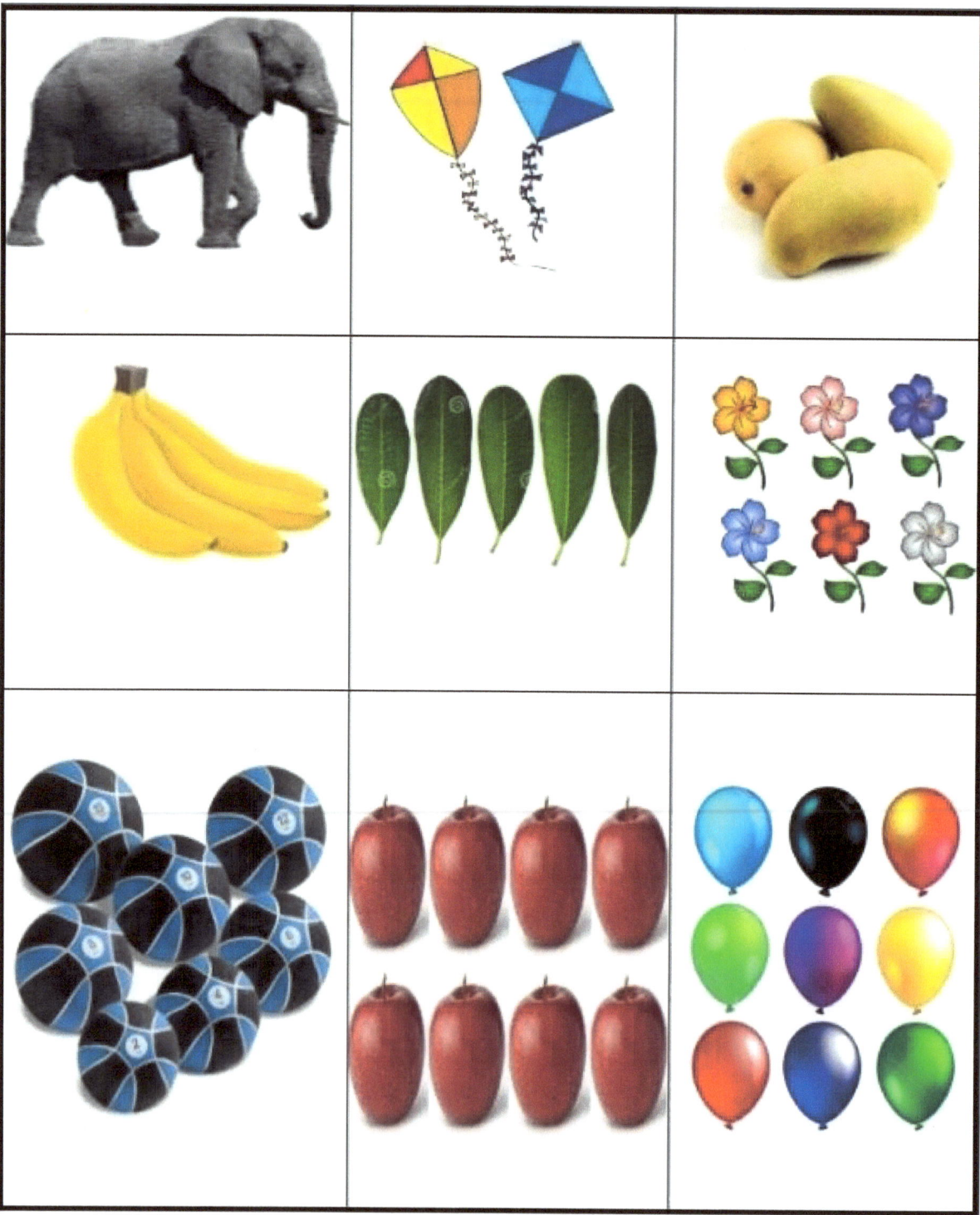

Multiplication Table

Lesson-1B

Let Us Learn From 0 to 9

0	Zero	5	Five
1	One	6	Six
2	Two	7	Seven
3	Three	8	Eight
4	Four	9	Nine

Lesson -1C
Let us learn the numbers from pictures.

0 means nothing..	1 bird
2 birds	3 deers
4 horses	5 bananas

Exercise-1A

Read aloud the number:

7	5	0	1	3
9	4	2	6	8

Read aloud the spellings:

Eight	Six	Three	Zero	Seven
Nine	Four	Two	One	Five

Writing Method

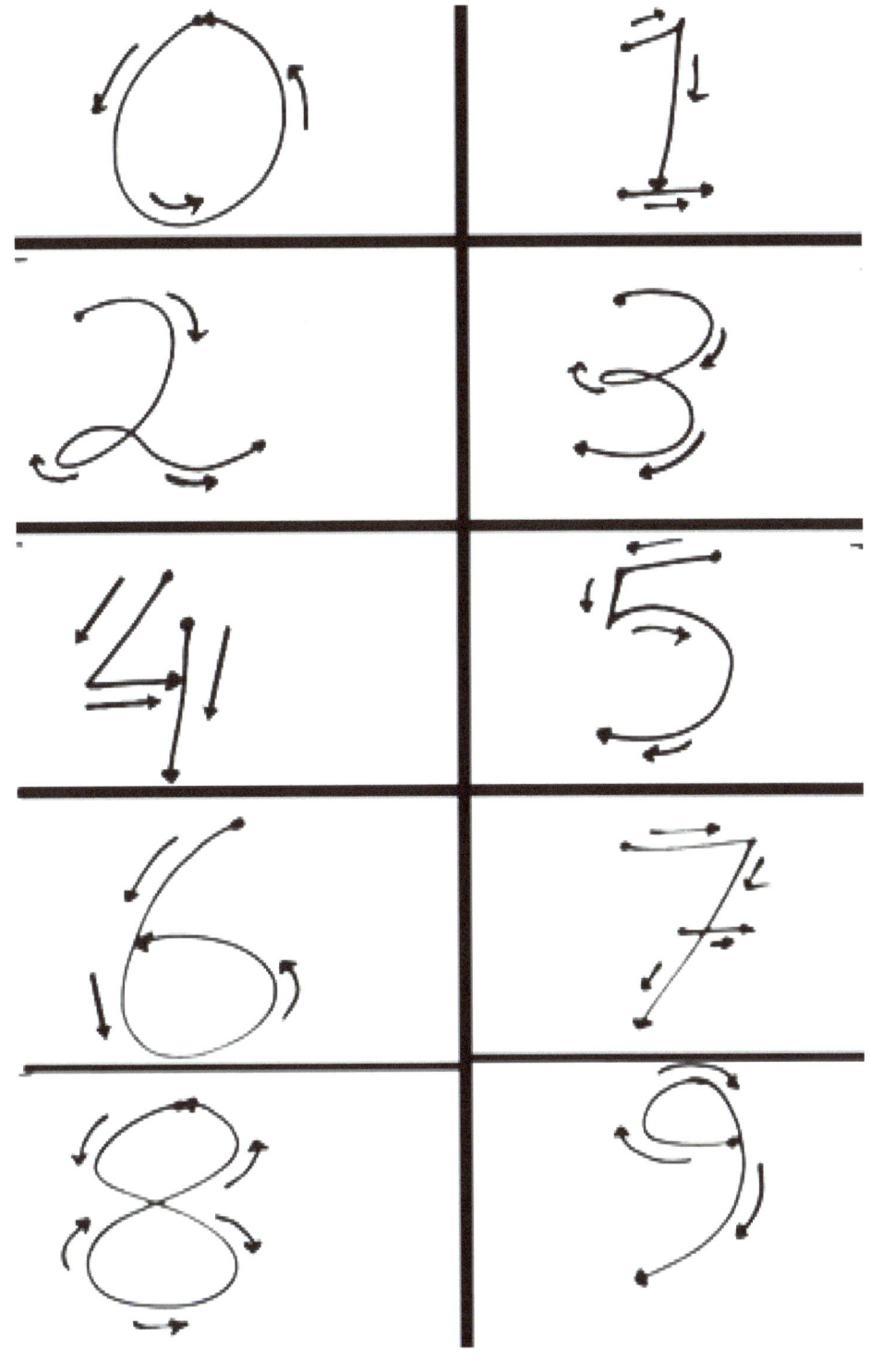

Multiplication Table

Exercise -1B

Let us count and write the numbers

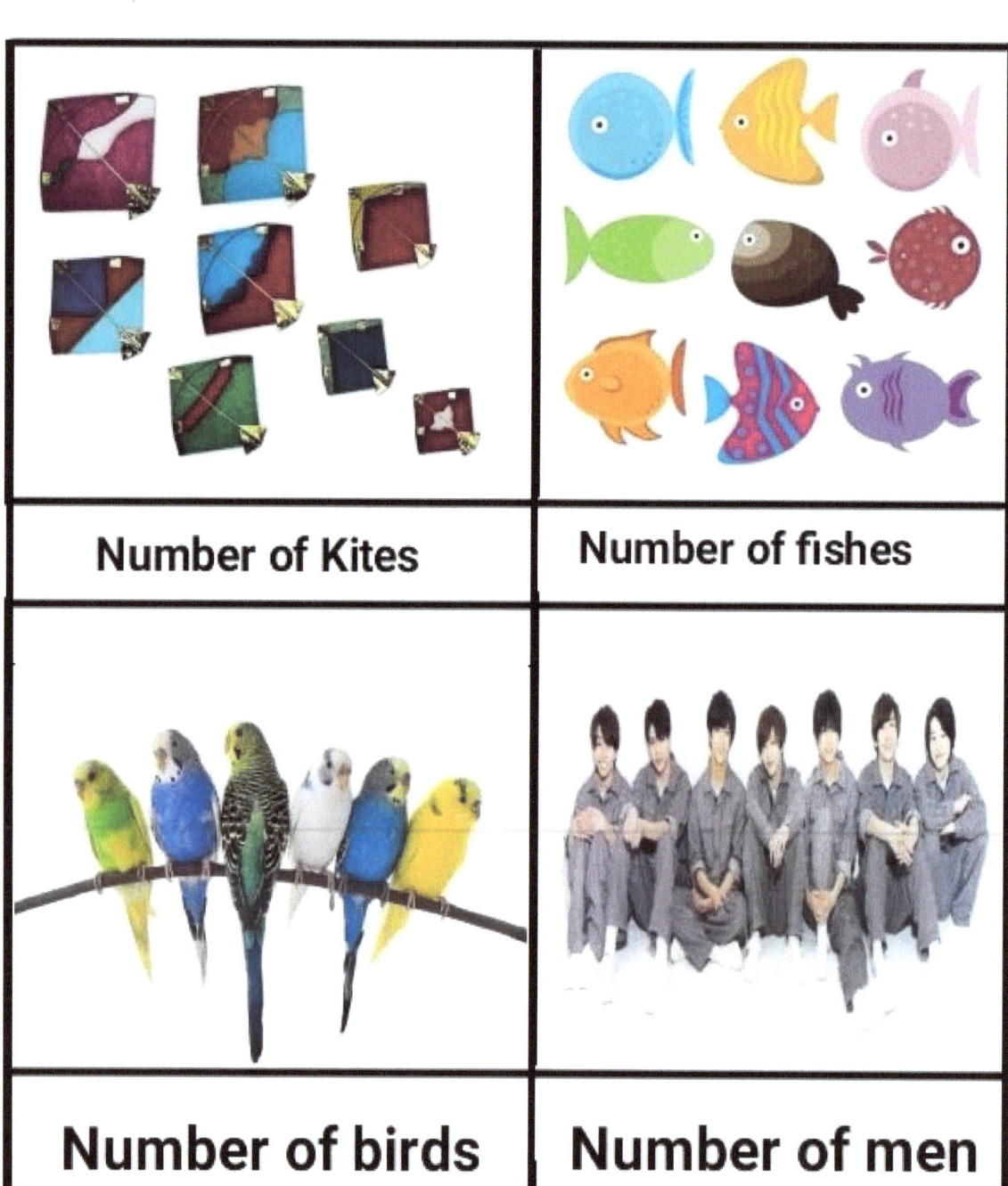

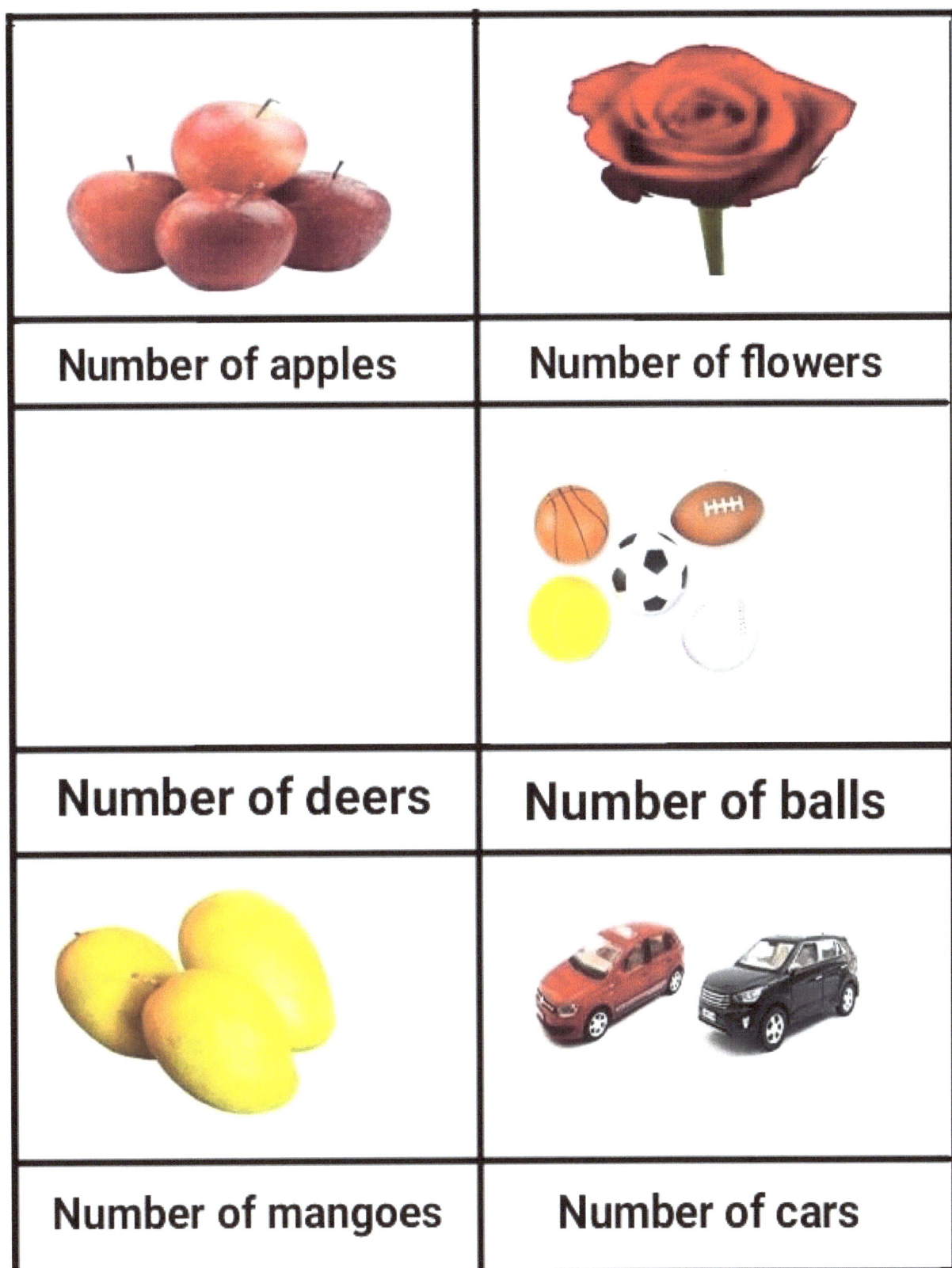

Multiplication Table

Exercise-1C

Write the spelling in the blank space:

8	3	6	9	0
7	5	1	4	2

Write the number in the blank space:

Eight	Six	Five	Three	Zero
Seven	Nine	Four	Two	One

Multiplication Table

Lesson-2

Let us learn Addition

Let us learn about addition. When two or more numbers are put together, then we get a result and the process is known as addition. Let us learn it in a practical way. Let us take few colourful blocks or pieces of stones. The blocks may be wooden or plastics, it does not matter. Already we have learnt to count and write from 0-9. Now take 1,2,3 i.e. 3 blocks in a box and 1,2 i.e. 2 blocks in another box. Now put all the blocks together in a box. And count. How many you get ? Count ! 1,2,3,4 and 5. Yes you get 5 blocks. See the figure below.

Now how to represent this it mathematically with paper and pen or on computer screen ?It is written as   3+ 2 = 5 '+' is known as 'plus' and '=' this is known as 'equals to' . So the whole equatioin is read as '3 plus 2 equals to 5'.

Same we can practice by taking 4 blocks and 3 blocks together and the result will be 7. See the figure below

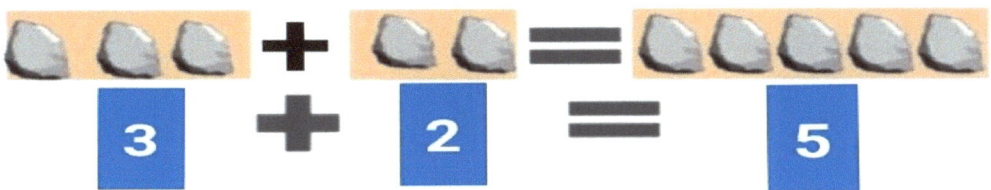

It is represented mathematically as 4+3=7 and is read as 4 plus 3 equals to 7.

It can also be written as 3+4=7   and then be read accordingly i.e. 3 plus 4 equals to 7.

Let us learn more addition in the described practical way by the following numbers and from Exercise - 2

7+1, 3+2, 6+2, 3+3, 7+7, 5+1

3+8, 2+1, 8+1, 5+1, 6+1

For the convenience of the students the addition table of all numbers from 0-9 is given in the table below. Because in this stage kids have learnt from 0-9, so only single digits nos are given. They will fill up the vacant boxes when they finish learning up to 20.

+	0	1	2	3	4	5	6	7	8	9
0	0	1	2	3	4	5	6	7	8	9
1	1	2	3	4	5	6	7	8	9	
2	2	3	4	5	6	7	8	9		
3	3	4	5	6	7	8	9			
4	4	5	6	7	8	9				
5	5	6	7	8	9	10				
6	6	7	8	9			12			
7	7	8	9					14		
8	8	9							16	
9	9									18

Exercise-2

Write the result of addition by following the practical methods described above.

2+3=	6+1=
4+2=	7+2=
9+0=	6+3=
1+7=	5+4=
5+0=	3+1=
0+6=	1+1=
0+0=	3+2=
3+3=	2+1=
3+4=	8+0=
5+1=	9=
6+2=	8=
8+1=	7=

Lesson-3

Let us learn from 10- 49

10	Ten	20	Twenty
11	Eleven	21	Twenty-one
12	Twelve	22	Twenty-two
13	Thirteen	23	Twenty-three
14	Fourteen	24	Twenty-four
15	Fifteen	25	Twenty-five
16	Sixteen	26	Twenty-six
17	Seventeen	27	Twenty-seven
18	Eighteen	28	Twenty-eight
19	Nineteen	29	Twenty-nine

Exercise-3

Read the numbers aloud and write the spelling in the copy book.

12	15	18	10	14
29	27	24	25	28
16	13	17	11	19
26	23	21	22	20

Read the spelling aloud and write the numbers in copy book.

Eleven	Thirteen	Nineteen	Ten	Fifteen
Twenty-eight	Twenty-three	Twenty-five	Twenty-one	Twenty-six
Eighteen	Fourteen	Seventeen	Sixteen	Twelve
Twenty-seven	Twenty-four	Twenty-nine	Twenty	Twenty-two

Multiplication Table

Lesson-4

Let us learn from 30-49

30	Thirty	40	Forty
31	Thirty-one	41	Forty-one
32	Thirty-two	42	Forty-two
33	Thirty-three	43	Forty-three
34	Thirty-four	44	Forty-four
35	Thirty-five	45	Forty-five
36	Thirty-six	46	Forty-six
37	Thirty-seven	47	Forty-seven
38	Thirty-eight	48	Forty-eight
39	Thirty-nine	49	Forty-nine

Multiplication Table

Exercise-4

Read the numbers aloud and write the spelling in the copy book.

34	33	32	31	37
48	41	43	45	47
38	36	30	39	35
40	42	49	44	46

Read the spelling aloud and write the numbers in copy book.

Thirty-eight	Thirty-one	Thirty-three	Thirty-two	Thirty-four
Forty-two	Forty-four	Forty-eight	Forty-three	Forty-nine
Thirty	Thirty-five	Thirty-nine	Thirty-six	Thirty-seven
Forty-one	Forty-five	Forty	Forty-six	Forty-seven

Multiplication Table

Lesson-5

Let us learn from 50-69

50	Fifty	60	Sixty
51	Fifty-one	61	Sixty-one
52	Fifty-two	62	Sixty-two
53	Fifty-three	63	Sixty-Three
54	Fifty-four	64	Sixty-Four
55	Fifty-five	65	Sixty-Five
56	Fifty-six	66	Sixty-Six
57	Fifty-seven	67	Sixty-Seven
58	Fifty-eight	68	Sixty-Eight
59	Fifty-nine	69	Sixty-Nine

Exercise-5

Read the numbers aloud and write the spelling in the copy book.

58	50	57	52	54
61	65	60	68	66
53	56	51	55	59
67	62	69	63	64

Read the spelling aloud and write the numbers in copy book.

Fifty-two	Fifty-four	Fifty-five	Fifty-one	Fifty-seven
Sixty-two	Sixty-Six	Sixty	Sixty-Nine	Sixty-Three
Fifty	Fifty-three	Fifty-six	Fifty-eight	Fifty-nine
Sixty-one	Sixty-Four	Sixty-Eight	Sixty-Five	Sixty-Seven

Lesson-6

Let us learn from 70-89

70	Seventy	80	Eighty
71	Seventy-One	81	Eighty-One
72	Seventy-Two	82	Eighty-Two
73	Seventy-Three	83	Eighty-Three
74	Seventy-Four	84	Eighty-Four
75	Seventy-Five	85	Eighty-Five
76	Seventy-Six	86	Eighty-Six
77	Seventy-Seven	87	Eighty-Seven
78	Seventy-Eight	88	Eighty-Eight
79	Seventy-Nine	89	Eighty-Nine

Exercise-6

Read the numbers aloud and write the spelling in the copy book.

78	76	79	70	77
81	85	80	86	82
73	75	74	72	71
83	87	84	89	88

Read the spelling aloud and write the numbers in the copy book.

Seventy	Seventy-Five	Seventy-One	Seventy-Eight	Seventy-Four
Eighty-Five	Eighty-One	Eighty-Seven	Eighty	Eighty-Four
Seventy-Seven	Seventy-Two	Seventy-Three	Seventy-Eight	Seventy-Six
Eighty-Two	Eighty-Six	Eighty-Nine	Eighty-Eight	Eighty-Three

Multiplication Table

Lesson-7

Let us learn from 90-99

90	Ninety
91	Ninety-One
92	Ninety-Two
93	Ninety-Three
94	Ninety-Four
95	Ninety-Five
96	Ninety-Six
97	Ninety-Seven
98	Ninety-Eight
99	Ninety-Nine

Exercise-7A

Read the numbers aloud and write the spelling in the copy book.

92	94	95	99	93
91	97	90	96	98

Read the spelling aloud and write the numbers in the copy book.

Ninety-Nine	Ninety-Three	Ninety-Four	Ninety-One	Ninety
Ninety-Six	Ninety-Five	Ninety-Seven	Ninety-Two	Ninety-Eight

Multiplication Table

Exercise-7B

In this stage students are advised to revise from Lesson 1 to Lesson 7.

Read the numbers aloud and write the spelling in the copy book. **Serially placed.**

0	10	20	30	40	50	60	70	80	90
1	11	21	31	41	51	61	71	81	91
2	12	22	32	42	52	62	72	82	92
3	13	23	33	43	53	63	73	83	93
4	14	24	34	44	54	64	74	84	94
5	15	25	35	45	55	65	75	85	95
6	16	26	36	46	56	66	76	86	96
7	17	27	37	47	57	67	77	87	97
8	18	28	38	48	58	68	78	88	98
9	19	29	39	49	59	69	79	89	99

Multiplication Table

Exercise-7C

Read the numbers aloud and write the spelling in the copy book.

Haphazardly placed.

87	67	54	45	34	37	25	8	32	44
51	22	63	69	27	31	53	82	99	50
71	35	0	64	85	7	33	72	68	49
4	42	61	57	38	19	90	21	55	73
66	93	16	5	11	28	46	96	23	83
77	47	91	13	94	74	43	59	20	81
36	1	24	75	98	9	62	40	65	48
56	2	15	39	29	30	92	3	58	18
10	14	6	26	17	79	86	80	97	89
70	88	12	76	95	41	52	60	78	84

Multiplication Table

Lesson-8

Let us learn bigger numbers

Dear kids so far we have learnt all two digit numbers from 0 to 99. Now we will learn about 3 digit numbers like 124 or 456 or more. Before that we will learn something new about two digit numbers. For example let us take the number 78. We generally read from left to right, so 7 comes first and 8 later. But if we read from right to left then 8 comes first and 7 later ! And that is how it is done while calculating the value of a number. As this number is 78. So, they are given a position and that is being calculated from right to left. So, it is regarded as '8' in the Unit position and '7' in the Tenth position. That signifies that the number is having 7 of 10's and one '8', so it is 78. It can be represented as below.

Tenth	Unit
7	8

We can try to understand the same by taking another number like 45. Here '5' is in the unit position and '4' is the 10 th position. So, it contains 4 of 10's and one 5. So, it is 45. See it below.

Tenth	Unit
4	5

But if the number consists of 3 digits ? So far we have learnt upto 99. What is after 99 ? It is 100. We take hundred as an example. It can be represented as below.

Hundredth	Tenth	Unit
1	0	0

Here is a third position and that is Hundredth's position. Here 1 is in the Hundredth position. So it is 100. If 2 is in the Hundredth position then it is two hundreds and so on. For example, take 156, it is being read as one hundred and fifty six. Take another example of 478. So, 4 is in the Hundredth position and it is read as 4 hundred seventy eitht.

And what in case of bigger numbers !! Say for example 5643. Now calculating friom righft to left the 4th position is named as House of Thousand and is occupied by 5. So, it is read as 5643 and can be represented as below :

Thousand	Hundredth	Tenth	Unit
5	6	4	3

In the same way 4789 can be read as Four Thousand Seven Hundred Eighty Nine and so on. Is not that awesome !! Now read aloud and learn the following numbers in the same manner. Take help of teachers.

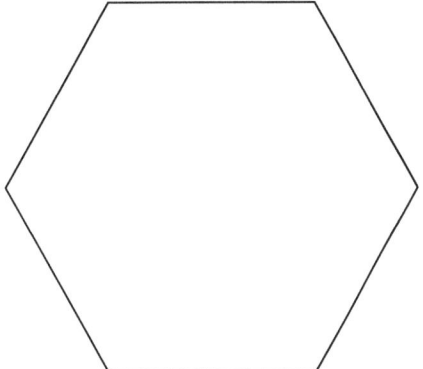

Exercise-8

Recognize the numbers, read aloud and write the spelling on copy book. Take help of teachers.

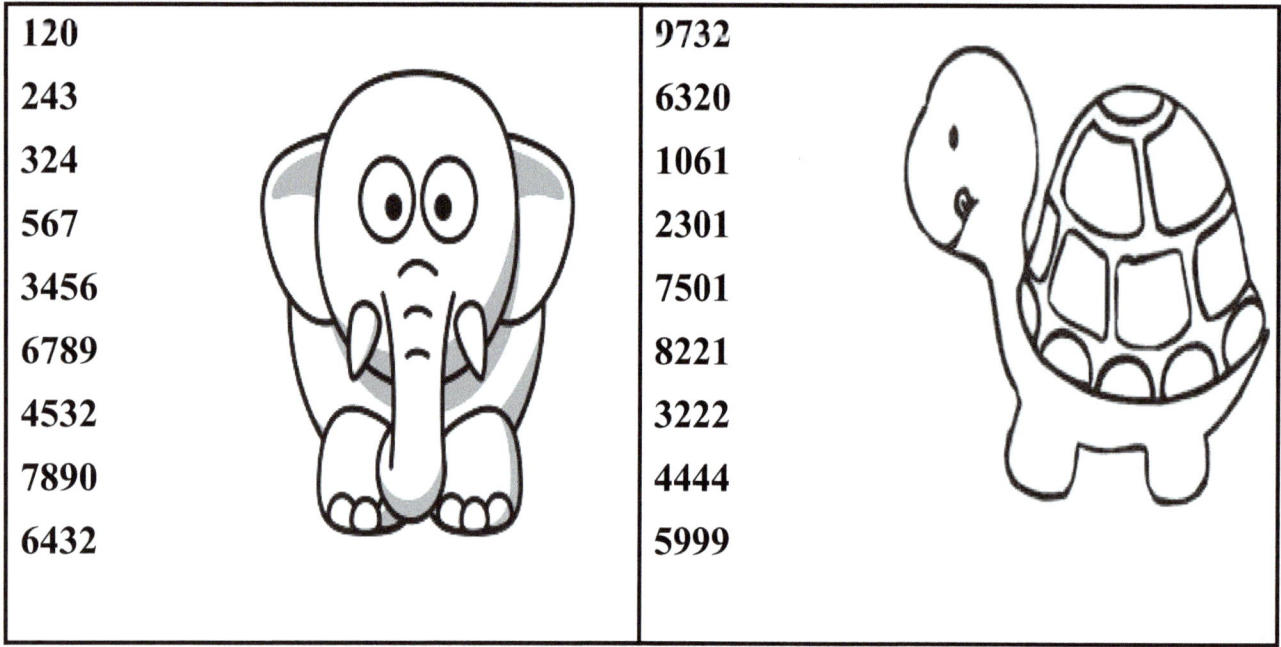

120	9732
243	6320
324	1061
567	2301
3456	7501
6789	8221
4532	3222
7890	4444
6432	5999

We will learn about bigger numbers in higher classes.

Lesson-9

Let us learn Multiplication

When a number is repeatedly added to itself, then it results in a product and the method is known as the process of multiplication. Let us learn it in a practical way. As described in lesson-2, let us manage some colourful wooden or plastic blocks or in absence of that pieces of stones. Now take 3 blocks and put them in a box. Again take 3 blocks and put in another box. . Now how many blocks you get ? Count. 1,2, 3,4,5 and 6. Yes you get 6 blocks. Now how to write in the copy book or computer ? It is written as :

3×2=6

'×' is read as into and '=" as equals to . So, the equation is read as 3 into 2 equals to 6. Here 6 is the product. 3 and 2 are multiplier and multiplicand.

Now if we take 4 blocks 3 times ? Then we get 12 blocks and is represented as

 4 ×3= 12.

 It is being read as 4 into 3 equals to 12

It can also be written as 3×4=12 and read accordingly i.e. 3 into 4 equals to 12.

Multiplication Table

Exercise-9

Let us solve the following problems by practical method. If needed take help of teachers.

8×7=	9×3=
6×7=	7×5=
9×6=	6×4=
3×2=	5×4=
4×2=	3×3=
4×5=	4×3=

Multiplication Table

Lesson-10

Multiplication is nothing but repeated addition

Yes Dear kids, multiplication is nothing but repeatation of addition. How is that ? Let us learn. Suppose you put 3 blocks 2 times in a box. So, you get 6 blocks and is represented as :

3×2=6

It can also be represented as   3+3=6 or   2+2+2=6

If you put 3 more blocks in it ? It becomes 9 blocks and is represented as :

3×3=9

It can be represented as below also:

3+3+3=9

If we add 3 more blocks to it ? Then it becomes 12 blocks and is represented as :

3×4=12

It can also be represented as :

3+3+3+3=12

So, we see if we express a particular number as multiplication, we get a short equation and if we express as addition, we get a long equation. For example,

4×5=20 This is short but…

4+4+4+4+4=20 or 5+5+5+5=20 These are long.

Similarly,

4×9=36 is short but…

4+4+4+4+4+4+4+4+4=36 or

9+9+9+9=36 These are long.

That is the reason that we follow multiplication process for bigger numbers.

Exercise-10

Write the following multiplication process as addition. One example is given below.

$3 \times 4 = 12$

Solution: $3+3+3+3=12$ or

$4+4+4=12$

Now do yourself.

$2 \times 3 = 6$	$6 \times 7 = 42$
$3 \times 5 = 15$	$7 \times 8 = 56$
$4 \times 5 = 20$	$3 \times 6 = 18$
$5 \times 5 = 25$	$4 \times 7 = 28$
$5 \times 6 = 30$	$3 \times 9 = 27$
$7 \times 5 = 35$	$5 \times 4 = 20$

Multiplication Table

Lesson-11
Let us learn the tables

Let us learn the tables. **Learning multiplication table is nothing but memorizing the product of multiplication of each two numbers.**

Generally in English the word 'Za' s being used or 'to' is used to read and memorise multiplication table. For example if we are learning multiplication table of 6 then it goes like :-

6 za 1 equals to 6

6 za 2 equals to 12

6 za 3 = 18 and so on…

Now it is the responsibility of the teachers to put it in the brains of kids very strongly that any number multiplied by 0 will result in 0. For example :

$1 \times 0 = 0$

$3 \times 0 = 0$

$4 \times 0 = 0$

Etc.

Again any number multiplied by 1 produces the same number. For example : -

$3 \times 1 = 3$

$5 \times 1 = 5$

$9 \times 1 = 9$

Etc.

And any number multiplied by 10 produces the same number with a zero on the right side. See the examples below :

$2 \times 10 = 20$

$3 \times 10 = 30$

$8 \times 10 = 80$

Same is the case of multiplication with 20. It is like multiplying with 2 and then adding a 0 on the right side. For example

$19 \times 20 = 380$ that means $19 \times 2 = 38$ and then placing a 0 on the right side.

So the tables of 0, 1, 10 and 20 is totally excluded from this book.

Only 3 tables are made for memorization. These are :Table-1, Table-2 and Table-3

However, below is given the traditional multiplication chart for convenience of teachers and parents. Hope that will help. But the students don't need to memorise that. They can read aloud. They are to memorise only the Table 1 and 2 and if possible Table-3. Table -1 is a must memorise for all students.

Multiplication Table

Traditional Multiplication Chart.

Table of 0	Table of 1	Table of 2
0 za 1 =0 0 za 2 =0 0 za 3 =0 0 za 4 =0 0 za 5 =0 0 za 6 =0 0 za 7 =0 0 za 8 =0 0 za 9 =0	1 za 1 =1 1 za 2 =2 1 za 3 =3 1 za 4 =4 1 za 5 =5 1 za 6 =6 1 za 7 =7 1 za 8 =8 1 za 9 =9	2 za 1 =2 2 za 2 =4 2 za 3 =6 2 za 4 =8 2 za 5 =10 2 za 6 =12 2 za 7 =14 2 za 8 =16 2 za 9 =18

Table of 3	Table of 4	Table of 5
3 za 1 =3 3 za 2 =6 3 za 3 =9 3 za 4 =12 3 za 5 =15 3 za 6 =18 3 za 7 =21 3 za 8 =24 3 za 9 =27	4 za 1 =4 4 za 2 =8 4 za 3 =12 4 za 4 =16 4 za 5 =20 4 za 6 =24 4 za 7 =28 4 za 8 =32 4 za 9 =36	5 za 1 =5 5 za 2 =10 5 za 3 =15 5 za 4 =20 5 za 5 =25 5 za 6 =30 5 za 7 =35 5 za 8 =40 5 za 9 =45

Multiplication Table

Table of 6	Table of 7
6 za 1 =6 6 za 2 =12 6 za 3 =18 6 za 4 =24 6 za 5 =30 6 za 6 =36 6 za 7 =42 6 za 8 =48 6 za 9 =54	7 za 1 =7 7 za 2 =14 7 za 3 =21 7 za 4 =28 7 za 5 =35 7 za 6 =42 7 za 7 =49 7 za 8 =56 7 za 9 =63
Table of 8	**Table of 9**
8 za 1 =8 8 za 2 =16 8 za 3 =24 8 za 4 =32 8 za 5 =40 8 za 6 = 48 8 za 7 =56 8 za 8 =64 8 za 9 =72	9 za 1 =9 9 za 2 =18 9 za 3 =27 9 za 4 =36 9 za 5 =45 9 za 6 = 54 9 za 7 =63 9 za 8 =72 9 za 9 =81

Table-1

Must be memorized. (2-9×2-9)

1	2	3	4	5	6	7	8	9
2	4	0	0	0	0	0	0	0
3	6	9	0	0	0	0	0	0
4	8	12	16	0	0	0	0	0
5	10	15	20	25	0	0	0	0
6	12	18	24	30	36	0	0	0
7	14	21	28	35	42	49	0	0
8	16	24	32	40	48	56	64	0
9	18	27	36	45	54	63	72	81

Multiplication Table

Exercise-11

Solve the problems with the help of Table-1

5×7=	8×8=	16=8×2	72=
8×9=	8×6=	4=	24=
4×6=	3×9=	12=	30=
3×7=	9×9=	21=	8=
8×5=	9×6=	42=	10=
7×5=	8×2=	36=	25=
9×5=	4×7=	18=	45=

Now how to learn from the tables? It is very easy. You are to find the intersect of column and row of the two involved numbers for which you are trying to find the result. For example if you are to find the product for 5×7 then choose either column or row for 5 or 7 and vice versa and see the intersecting cell and that is your desired result and in this case it is 35. And for memorization one can follow either vertical or horizontal way. Vertical way is recommended.

Lesson-12

Let us learn multiplication of two digit numbers.

Now we will learn the multiplication of two digit number like 12 or 19. Here 11-19 are multiplied by 2-9. Memorisation is suggested. So let us start.

Table-2

1	11	12	13	14	15	16	17	18	19
2	22	24	26	28	30	32	34	36	38
3	33	36	39	42	45	48	51	54	57
4	44	48	52	56	60	64	68	72	76
5	55	60	65	70	75	80	85	90	95
6	66	72	78	84	90	96	102	108	114
7	77	84	91	98	105	112	119	126	133
8	88	96	104	112	120	128	136	144	152
9	99	108	117	126	135	144	153	162	171

Multiplication Table

Exercise-12

Solve the problems with the help of Table-2

5×13=	15×5=	105=	102=
7×19=	12×7=	135=	39=
11×2=	17×7=	171=	28=
12×3=	12×9=	65=	64=
19×5=	16×8=	136=	60=
15×6=	17×9=	152=	78=
14×5=	18×9=	119=	104=

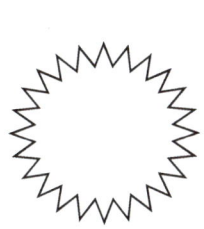

Lesson-13

Let us learn multiplication of Bigger numbers

11-19 x 11-19

Multiplication Table

Table-3

×	0	11	12	13	14	15	16	17	18	19
11		121	0	0	0	0	0	0	0	0
12		132	144	0	0	0	0	0	0	0
13		143	156	169	0	0	0	0	0	0
14		154	168	182	196	0	0	0	0	0
15		165	180	195	210	225	0	0	0	0
16		176	192	208	224	240	256	0	0	0
17		187	204	221	238	255	272	289	0	0
18		198	216	234	252	270	288	306	324	0
19		209	228	247	266	285	304	323	342	361

Multiplication Table

Exercise-13A

Solve the problems with the help of Table-3

18×13=	324=
13 × 11=	228=
14 × 11=	304=
15 × 14=	323=
16 × 13 =	361=
18 × 17=	195=
13 × 13=	192=
19 × 18=	204=
18 × 14=	221=
16 × 11=	255=
14 × 13=	121=
12× 11=	225=
14 × 14 =	165=
18 × 15=	187=
17 × 17=	209=

Multiplication Table

Exercise-13B

Solve the problems with the help of Table-1,2 and 3

6 × 6=	117=
7 × 9=	108=
5 × 6=	114=
3 × 17=	144=
5 × 19=	57=
16 × 19=	24=
13 ×16=	27=
12 × 3=	224=
19 × 2=	272=
14 × 2=	238=
15 × 2=	285=
13 × 2=	216=
18 × 16=	247=
16 × 13 =	324=
18 × 11=	288=

Teachers or students can make their own tables to practice more.

Extra 3 pages given to keep notes.

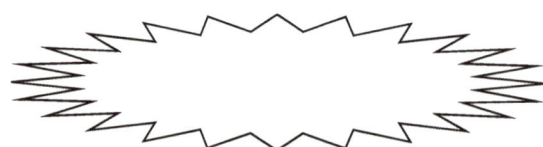

Multiplication Table

Multiplication Table

This book is available in Flipkart, Amazon and notionpress.com
This book has also got a black and white version with lower prices.
E-book is also available in Amazon KINDLE STORE.

www.ingramcontent.com/pod-product-compliance
Lightning Source LLC
Chambersburg PA
CBHW051217220526
45473CB00003B/1073